Fr Schröder

The Homoeopathic Poultry Physician (poultry veterinarian)

Or, plain directions for the homoeopathic treatment of the most common

ailments of fowls, ducks, geese, turkeys and pigeons.

Fr Schröder

The Homoeopathic Poultry Physician (poultry veterinarian)
Or, plain directions for the homoeopathic treatment of the most common ailments of fowls, ducks, geese, turkeys and pigeons.

ISBN/EAN: 9783337291969

Printed in Europe, USA, Canada, Australia, Japan

Cover: Foto ©berggeist007 / pixelio.de

More available books at **www.hansebooks.com**

THE

HOMŒOPATHIC

POULTRY PHYSICIAN

(POULTRY VETERINARIAN);

OR,

PLAIN DIRECTIONS FOR THE HOMŒOPATHIC TREAT-
MENT OF THE MOST COMMON AILMENTS OF

Fowls, Ducks, Geese, Turkeys and Pigeons.

BASED ON THE AUTHOR'S LARGE EXPERIENCE, AND
COMPILED FROM THE MOST RELIABLE
SOURCES,

BY

DR. FR. SCHRÖTER.

Translated from the German.

BOERICKE & TAFEL,

NEW YORK:
145 GRAND STREET.

PHILADELPHIA:
635 ARCH STREET.

1880.

PREFACE.

As yet there existed no work on the treatment of the numerous ailments peculiar to poultry by Homœopathic remedies.

This fact and the numerous inquiries from farmers and bird fanciers, induced the publishers to have a translation made of this little book from the German, they having imported and disposed of many hundred copies of the original in the last few years.

It embodies, in a concise form, everything known on the subject, and practical and common-sense advice given in addition to the Homœopathic treatment, will be heartily appreciated by every intelligent raiser of poultry.

Those faithfully following the plain directions given, will be surprised how speedily and easily hitherto untractable poultry diseases may be overcome, and become convinced of the superiority of the "mild treatment" over that hitherto followed or recommended.

THE PUBLISHERS.

CONTENTS.

THE HOMŒOPATHIC
POULTRY PHYSICIAN.

PART I.

Importance of Homœopathy and how to apply it to the cure of Diseased Poultry.

INTRODUCTION.

THE importance of the homœopathic method of cure, discovered by Dr. Sam'l Christ. Fr. Hahnemann,* for the treatment of the diseases of our domestic animals, is every year more and more recognized, and the opposition to it is steadily diminishing. When now and then

---

* Hahnemann was born at Meissen, Saxony, April 10, 1755, and died at Paris, July 2, 1843.

some zealot is heard yet against Homœopathy, it is either because of his ignorance of its essence, or because some private interest prevents him from doing honor to truth. Homœopathy, however, in spite of all infestations, principally those of former times, has triumphantly run its course, and now occupies, at the side of Allopathy, a position not only worthy of itself, but as we may safely portend, far superior to its antagonist. And this by right! Even the most simple farmer, suppose he has correctly diagnosed the disease of his animals, is enabled to cure them rapidly, easily, and with little expense, by the aid of Homœopathy. This is an inestimable advantage in places often situated many miles from a veterinary physician, where it would sometimes take days before the assistance of a competent healer could be procured. But it is well known, that many of the diseases of animals, if help be tarrying, are apt to soon take so stubborn a character that in many cases it will take a longer time and cause much more trouble and, therefore, much greater expense to effect a cure, while in others the course of the disease has become fatal and no more help can be given. It is evident, that this may subject the proprietor to great losses. On the contrary, it is a matter of experience that even the most virulent diseases, as for instance colic, can with ease and without risk be removed by a homœopathic remedy if administered in time.*

This is also confirmed by the opinion of an agricultural authority, Councillor Kleemann, who says, that since the treatment of animals by homœopathy, scarcely one-half of them have succumbed; whilst formerly, under allopathic

---

* As characteristic in this respect, we may mention a case lately recorded in the Monthly of the Farmers' Association of the Provinces of Brandenburg and Nether-Lusatia: "A veterinary doctor had lost his own horse by colic. Although he knew well that his horse could be saved if he would accept some homœopathic remedy from me, he preferred to see his horse die to resorting to hateful homœopathy. When, however, the horse which had been substituted for the fallen one was also attacked by colic, the anxiety that this horse might be lost too, overcame the hatred of homœopathy. The Doctor cantered into my yard in hot haste, and showed his horse to me, exclaiming: 'My new horse has colic also; if you don't help, I will be lost, for my art is at an end.' I gave the horse a small dose, it perfectly recovered, and the very same day the Doctor returned to express his thanks to me."

treatment in heavy epidemics, rarely one was saved. But this, indeed, can cause no wonder with any one who knows that the homœopathic method alone is in accordance with nature, and therefore the only reliable one.

This, however, has its deep foundation in the spirit of homœopathy which consists in *curing like through like;* that is, in other words, in applying to a given disease a remedy which, if taken in larger doses, would produce a similar disease in the system to the one the patient is afflicted with.

Has not this principle its thousand-fold confirmation in daily life? The heated day laborer, before slaking his burning thirst with cold beer or water, takes, even in the most scorching sun, first a warming sip of whisky, which does him good; the experienced cook does not put her scalded hand into cold water, which would cause blisters, but she holds the injured limb near the fire whereby the heat is extracted; the saving housewife does not thaw her frozen potatoes or apples in hot water, which would make them rot, but puts them in ice-cold water, thus restoring them to an eatable condition; even the allopathic physician, for the restoration of frozen men, does not order the application of heat, but of ice-cold snow in frequent repetition. We see therefrom, that the spirit of homœopathy is deeply founded in Nature, and is the cause of its certain effect by the *most simple remedies.*

This, however, is one of the great advantages Homœopathy has over Allopathy. Formerly the old-school veterinary physician ordered for the diseased animal some draught brewed together of ten different ingredients, and then he wondered if the disease, instead of better, got worse. Now, in our days, as cannot be denied, Allopathy has learned something from Homœopathy, for though those old tapeworm recipes have been shortened, yet, nevertheless, a prescription of but one simple remedy, as Homœopathy administers, is still a rare thing. Whoever will cast a glance into the so-called allopathic "Cure-books" will find that what we just said is absolutely true. Homœopathy, on the contrary, prescribes but *one* remedy, the action of which has been accurately scrutinized and defined, and in a few cases only resorts to alternating with two remedies. A second, but not lesser advantage, is the *cheapness* of the homœopathic treatment. Three or four vials of the homœopathic remedies scarcely cost more than one glass of any old-school medicine. Moreover, the former if properly kept, will remain unspoiled and available for years, and can easily be re-supplied, which cannot be said of the allopathic remedies. Nay, one single vial of Aconite, Arsenicum, etc., which costs not more than a quarter, can cure hundreds of cases.

A further advantage is evidently the *easy way of administering* the homœopathic medicine. We will in this respect only refer to the great trouble in administering allopathic medicines to the hogs. How often is main force required to break their snouts open, and the application of gags in order to forcibly pour in the medicine, whereby often the animal is injured by its fierce struggling and worrying. The action of the medicine in such cases can be but insignificant or equal to naught. And now imagine how easily and willingly the hog takes homœopathic medicine, particularly if administered in milk. It is really distressing to see animals worried and tortured to take draughts and pills which they refuse, whilst the same end may be attained so much more simply and easily.

All these advantages of homœopathy so apparent in the treatment of our larger domestic animals, appear much more eminently if taken into consideration for the treatment of our diseased lesser domestics, as in the first place, the *poultry.* While Allopathy often looks hopelessly and helplessly at a diseased fowl, shaking her head gravely but unable to give any relief, Homœopathy, just in critical cases, affords help surely and quickly, and often astonishingly, with one single dose.

Diet and Nursing of the Animals during Disease.

The *diet* of diseased animals during homœopathic treatment must not be overlooked, since upon it success depends to a great extent. In the first place, the proprietor has to take care to keep off from the sick animal all those influences which foster the diseased condition, in order to enable the vital powers to receive the action of the medicine in all its purity. Principal care has to be taken, therefore, to prevent all collateral allopathic quackery; also to remove from the fodder of the diseased animal everything spicy, as strong-smelling herbs, seeds, etc. It is also necessary not to give it any food or drink at least half an hour after, and if possible, before administering medicine. The patient has to be separated from the healthy animals, and put in a place which is very dry and clean and free from strong odors.

We need scarcely to remark that by treating diseased animals kindly we will sooner and better restore them to health than by acting roughly and impatiently. Although this is already enjoined by the charity which man owes to the smallest worm, our own interest much more commands us to deal with sick animals in a loving and sparing way, since the opposite treatment may easily cause an aggravation of the disease.

Choice of the Homœopathic Remedy.

In order to choose with a degree of certainty the correct remedy for a disease, it is necessary to first accurately ascertain the diseased condition of the animal, to compare it with the normal (healthy) condition, and so to establish every deviation from the latter as a disturbance in the organism that is as a symptom of disease. In this way there will soon appear the distinction between *principal symptoms*, which are characteristic of the present case, and *accessory symptoms*, which are but of a transitory nature. And this will enable us to choose the remedy only according to the principal symptoms governing the present emergency. If, in this choice, an error should be committed, which will be known by the non-appearance of a speedy favorable turn in the disease, this would be of no great account, as the blunder is not followed by any disadvantage, and there remains only to search for another remedy, based on a quiet and careful examination and comparison of the diseased condition.

The alternate use of two different homœo-pathic remedies is, as a rule, only then recommendable if both seem to suit the case equally well, but neither to hit the actual case squarely. To change the medicine, that is, to prescribe another remedy, is commanded when the disease presents so altered an aspect as to imperatively call for a different remedy. As to which remedy has in such a case to take the place of the former, we have to be entirely guided by the character of the case. A change in the medicine becomes necessary also, if we observe that a remedy, after having worked for some time, shows no more effect. In chronic diseases, however, it is advisable to watch the effect for a greater length of time, and in some cases to even make a longer intermission on purpose.

On the Repetition of Homœopathic Medicines.

As a rule, *one single dose* of the remedy will *not* be sufficient to remove the disease, and we will, therefore, have to repeat it. In quite superficial ailments only one single dose may suffice. Whenever the administered remedy shows no effect at all, it is, as we have already said, a sign that a wrong, not fitting remedy has been selected, and some other one more appropriate has to be applied. If the remedy, however, acts, but the improvement stops suddenly, the medicine has to be given repeatedly, and in acute diseases oftener than in chronic complaints.

Any aggravation of the disease immediately after the administration of a remedy—the so-called *primary* action—is usually a sure indication of the future healing effect. Such an aggravation must, therefore, not cause any trouble or induce to apply a different remedy, or to prematurely repeat the well-selected medicine.

The Dose of Homœopathic Medicines.

The dose of the homœopathic medicines has to be guided partly by the species of the animals, partly by the character of the disease, partly by the remedy itself. Older animals, for instance, require larger doses than younger ones;

as well diseases of a rapid course call for such
more than slow ones. To younger animals
therefore a dose of 1 to 2 pellets, to more ad-
vanced ones of 2 to 5 pellets, to older of 4 to 7
pellets of the homœopathic remedy ought to be
given, though on the whole it is advisable to
rather give a couple of pellets more than too
few, as a surplus will never do harm and be
soon equalized by nature, which, however, is
not intended to mean that "much helps much,"
and may not possibly do harm.

A further rule, based on the nature of disease,
is: that in diseases of *rapid course* (acute), the
medicine has to be administered in *shorter*, in
ailments of *slow course* (chronic) in *longer* in-
tervals. In acute cases, therefore, according to
their vehemence, 4, 6, and 8 doses have to be
given, and in very severe cases a dose every
hour or even every half hour may be required.
In chronic complaints *two* or *one* dose a day,
and sometimes not more than one a week will
suffice.

It is best to administer the medicine in solution, by dissolving a number of pellets in a

glass of water or milk and giving a spoonful for a dose.

The administration is effected by securing the sick fowl on one's lap with the left elbow, seizing its head with the left hand, and opening its bill with the left thumb and forefinger, at the same time attending the neck of the animal and keeping its head up. In this position the medicine from the spoon, held by the right hand, is slowly poured into the throat of the animal, taking care of its being swallowed down by the patient; although it may be remarked, that to swallow the medicine is not even indispensable, as this will act by its mere contact with the mucous membranes; nay, the very vapor of the medicine alone will affect the nerves of the animal, and influence the vital force in the mildest but most powerful manner. This, for instance, is of incalculable value in the treatment of the bees, since it needs no proof that to them no remedy could be administered in the usual way.

In *external* injuries, as contusions, sprains, dislocations, etc., and generally in all unbloody lesions, the so-called *tinctures*, principally Arnica montana, Urtica urens, Thuja occidentalis, Symphytum officinale, are applied externally and *undiluted*. Their external application alone is, however, not to be recommended for the reason that every external ailment, insignificant as it may be in itself, affects the inward organism. The internal administration of the appropriate remedy has therefore to be combined with its external use. For *bleeding wounds*, however, a teaspoonful of the pure tincture has to be poured into a cupful of pure spring or rain water, and this solution, well-stirred, to be applied in the form of poultices or bathing, etc.

How to preserve Homœopathic Medicines.

Those homœopathic medicines mentioned in this pamphlet ought to be procured in time and always kept in store, in order to have them on hand in case of need. The greatest attention has also to be paid to their preservation, so as to protect them from all injurious influences which might tend to infringe their power and to neutralize their wished-for efficiency in diseases. All the remedies must, therefore, be kept in a dry, cool place, and where strong odors, as the exhalations of strong-smelling flowers, roasted coffee, tobacco fumes, vapors of sulphur, etc., have no access. Every vial of medicine should, immediately after having been used, be well closed again with the same stopper; for if one were to transfer the stopper of one vial to another, all the medicine of the vials, the stoppers of which had been exchanged, may be spoiled, be unreliable, be rendered unfit for further use. While, if these precautions are rigidly observed, the medicines will preserve their efficacy for years.

The most important and most frequently used homœopathic remedies are the following:

Absinthium, wormwood; Acidum sulphuricum, sulphuric acid; Aconitum napellus, aconite, wolf's-bane, monk's-hood; Arnica, arnica montana; Arsenicum album, white arsenic; Belladonna, deadly nightshade; Bryonia alba, begony; Chamomilla, chamomile; China, peruvian bark; Cina, worm-seed; Dulcamara, bittersweet; Euphrasia, eyebright; Hyoscyamus niger, henbane; Ipecacuanha, ipecac; Lycopodium, wolf's-foot; Mercurius vivus, quicksilver; Nux vomica; Opium, white-poppy juice; Pulsatilla, meadow anemone; Rhus toxicodendron, poison sumach; Spongia marina tosta, roasted sea-sponge; Staphisagria, St. Stephen's herb; Sulphur, brimstone; Thuja, arbor vitæ; Veratrum album, white hellebore; Tinctura arnicæ, arnica tincture; Tinctura Symph., tincture of Symphytum.

Besides the above-named drugs a great many others are indeed used by homœopathic physicians and veterinarians; the above-named, however, are, as we said before, the most important and most frequently used. Every proprietor of animals, therefore, ought to constantly have them on hand, and it is of importance that the greatest caution be exercised in procuring them. In order to be sure of their good and strong quality they ought never to be ordered but from homœopathic pharmacies of good standing. A medicine case holding thirty of the mostly used remedies mentioned in this little book, in vials holding over fifty doses each, will be furnished for five dollars, including book by the Publishers.

PART II.

The Diseases of the Poultry and their Homœopathic Treatment.

---

a. Introduction.

OF the poultry, under which appellation we comprise *pigeons, chickens, geese, ducks,* and *turkeys,* some old-fashioned proverb says: "Whoever wants to become a pauper, and don't know how, let him keep much poultry." But this proverb comes from some old crony's rubbish-box, and whoever invented it was a poor calculator or a malicious one. Experience, on the contrary, has proved in a most striking manner that the proverb should rather read, "Whoever wants to become a rich man, and don't know how, let him keep much poultry." For the fact is that poultry-culture (farming) if carried on judiciously and carefully, yields the largest net profits of all the agricultural branches with comparatively the smallest expense and risk, suiting at the same time the small farmer as well as the extensive. To prove this, however, would overtax the narrow limits of this pamphlet.

The profitableness of poultry farming is particularly well known in France and Belgium, where this branch of the agricultural business is highly flourishing. The fattened and truffled turkey, capons and chickens (pullets) of France, are world-famous, and form an important article of export, and the source of an income of many thousands of dollars a year to that country.

Of course if poultry-raising is carried on with want of knowledge, judgment, and industry, it may be unprofitable, and especially so if the farmer is ignorant in regard to the knowledge and cure of the diseases of the poultry. Every poultry-raiser ought, therefore, to be a *poultry-physician* too. What, for instance, were the use of a chicken-yard of a hundred or more fowls, if a third or even one half of them should yearly be lost by some fatal epidemic? In such a case the profit would certainly be small. The poultry, however, is liable to a great many diseases, some of which not seldom assume the form of fatal epidemics, and there are many instances of costly specimens of proud outlandish roosters, stately hens, precious pigeons or already fattened turkeys having been lost, because the proprietor did not understand how to cure the diseased animal. Although it may be true in general that the diseases of our tame fowls in most cases are caused by heat or cold, by either too copious and stimulating or too insufficient, spoiled, and inappropriate food, by want of proper exercise in the open air, by uncleanliness, by poisoning or external injuries, by noxious climatic influences, etc., and that these causes may be avoided by better attention and care being given to the poultry, it can still not be denied that there are diseases which will befall the fowls in spite of all precaution. These are the reasons for discussing in the following chapters the diseases of the poultry in a thorough manner, and for stating the best-proved homœopathic medicines for their relief.

b. A Glance at the Natural History of the Pigeon and its Diseases and Homœopathic Treatment.

We are justified in assuming that all the different varieties of the pigeon, of which there may be thousands, have their common origin from the field-pigeon, which belongs to the order of Gallinaceæ and the family of "round-flyers," Gyrators. It is quite as certain that the pigeon is an indigenous wild-fowl of Southern Europe and Middle Asia. At the shores of the Mediterranean it is still found wild in numerous flocks. Its tendency to turn back into wildness, in which it selects edifices and ruins for its abode, is generally known.

From Brehm's "Natural History and Culture of the Pigeon," we clip the following notes:

"All pigeons have a close plumage, consisting of somewhat harsh feathers, pointed wings with hard flag-feathers, and a middling long or somewhat stretched body. They fly very swiftly and with noise; they migrate regularly and pretty far and in large flocks in the moderate climates, are wild, very cautious, and shy, but tamable and generally popular for their beauty and grace. They feed, with preference, on seeds, especially fragrant and sweet ones, as aniseed, etc., and feed their young from their crop, which at this time acquires the peculiar quality of transforming the half-digested seeds into a cheeselike substance, which is appropiate to the weak digestion of the newly-fledged young, and continues until they are fully fledged. The female is a little smaller than the male, which alternates with the former in hatching, and assists her in bringing up the young. All species live in monogamy."

Pigeons, however, are not only raised for their grace and beauty, but also on account of their flesh. The flesh of a young pigeon is particularly juicy, and furnishes a palatable roast. A good fleshy, young pigeon, as all sweet-teeth assure us, gives the daintiest roast of all fowls. The flesh of older pigeons, on the contrary, is somewhat tough, but, if minced, very fitting for pastry filling and strong broths.

The pigeon is subject to the following diseases:

1. SMALL-POX.

SYMPTOMS.—This is an eruptive disease accompanied by high fever and characterized by pearl-like pustules, surrounded by inflammatory redness. In the pustules an infectious stuff is generated which often spreads over a whole flock, and in this way sometimes makes great havoc. With the old pigeons these pustules may already appear at the base of the bill, whilst with the young ones they appear principally at the ears and beneath the wings.

CAUSES.—This malady we may suppose most frequently to be caused by unfavorable atmospheric conditions. Another cause is its infectious character, the propagation being easily effected by the immediate contact of some pustulous animal, communicating some of the pustular liquid to the skin of the healthy bird. But by the mere exhalation of diseased animals the malady may also be spread, so that, as already mentioned, often a whole flock may be affected.

TREATMENT.—**Arsenicum** and **Rhus toxi-**

codendron, 2 or 3 pellets dissolved in water, and a dose given every two hours in alternation.

In addition the diseased fowls must be separated from the healthy, in order to protect those from infection.

Pustulous pigeons are not to be used in the kitchen.

2. CANARY PLAGUE.

SYMPTOMS.—The stomach is greatly swelled, the animal quiet and sad. This disease befalls the female only.

CAUSE.—Too great heat on the part of the male bird.

TREATMENT.—**Aconite** twice a day has done good service in such cases. But this malady having its cause in the physical conformation of the female bird, there is no better remedy than to give her a more modest partner.

3. EMACIATION.
(Consumption Roup).

SYMPTOMS.—The pigeons affected with this malady emaciate visibly, the breast-bone often sticking out like a sharp knife. This symptom

is often accompanied by great debility, constipation, or diarrhœa.

CAUSES.—This malady has its foundation in
some inward defect, weak digestion, marasmus,
etc.

TREATMENT.—The principal remedies are
Arsenicum and **China** in alteration. **Nux
vomica** ought to be given where there is constipation, **Pulsatilla** for diarrhœa.

DOSE.—2 to 3 pellets in water three times a
day.

4. DIARRHŒA.

(Cholera).

SYMPTOMS.—They appear in the evacuations
(fæces, dung) of the animals, these being of a
very liquid quality.

CAUSES.—This complaint is usually caused
by too copious but unwholesome food. It may
also be owing to cold in the stomach in cold,
damp weather, or have its foundation in some
inward disease.

TREATMENT.—The principal remedy is **Ipecac.** If unwholesome food be the exciting cause

Arsenicum is indicated. **Chamomilla** too has answered well.

Dose.—As above.

5. EPILEPSY.

Symptoms.—This ailment shows itself by a quite peculiar and remarkable contortion of the throat-muscles, and is more frequent with the female than the male bird. If the birds be touched in this condition they give signs of being in severe pain.

Causes.—This malady is based on a morbid condition of the nervous system.

Treatment.—In a first attack of this kind **Aconite** and then **Stramonium** may be given. Against the fully-developed disease there is no remedy.

Dose.—2 to 3 pellets in the morning and evening.

6. ULCERATED THROAT.

Symptoms.—The ulcers usually consist in some fleshy excrescence in the throat, which grows to an extent to threaten suffocation.

2

CAUSES.—These are to be found in an abnormal formation of the cellular tissue.

TREATMENT.—As generally useful in this malady **Aconite** and **Bryonia, Arsenicum** and **Sulphur,** may be mentioned; 2 or 3 pellets of either in a little water every two hours. A cure, however, is difficult. The most hopeful proceeding is to cut the ulcer out with scissors, and to cauterize its roots as soon as it makes its appearance. If it should recur, however, the pigeon is lost without help.

7. FRACTURE OF BONES.

SYMPTOMS.—A swelling appears over the fractured bone, which, if examined, causes the animal great pain. A fracture of the thigh-bone is easily to be ascertained by the animal not being able to rest on the broken leg, this appearing, as it were, as transfixed at the broken place, while by examining this it will show some pliability as of an articulation where there ought to be none. If the continuity of the bone be broken, the two ends often overlap each other, so that one can feel them protruding or even splinters of the bone, and the limb will appear shortened. Soon the fractured place will become inflamed and surrounded by painful swelling.

CAUSES.—Such fractures are very frequent as the consequence of a thrust, stroke, blow, etc.

TREATMENT.—The fractured bone must first be brought into its natural position; the injured part has then to be dressed with linen straps, over which a couple of grooved splints are applied, and this bandage frequently moistened with a lotion of Symphytum, one part to eight or nine of water. After a week the correct position of the bones has to be examined, and the bandage renewed and the lotion applied in the same manner. This procedure has to be repeated several times till the complete restoration of the animal. It is evident that during the treatment the patient has to be separated from the healthy animals and kept in a sequestered place, and that on the whole this treatment requires great caution and care on the part of the owner.

8. LICE.

SYMPTOMS.—Little black lice, quite different from the common pigeon-lice, are to be found in clusters on young pigeons, whom they kill by draining. They gather particularly below the wings. Leopold in his "Pigeon-Fancier" says about this subject: "In fact pigeons are almost always possessed of lice, fleas, and even bedbugs, in greater or lesser quantity, but the case we treat of here is quite different from that." Whether these parasites be really lice or mites (Acarus) as Brehm thinks, is of no account here.

CAUSES.—This plague commonly sets in if in very dry summers the old pigeons have no water for bathing in, and their cotes are not often enough cleaned.

TREATMENT.—Keep the dove-cotes and nests as clean as possible, and provide for sufficient water for bathing in dry summers. To the birds suffering from lice, **Sulphur,** and when there is great debility **China,** a dose three times a day, may be given. It has done good service also to sprinkle the floor of the cotes with camphorated water.

9. MOULTING.

(Annual feather-shedding.)

SYMPTOMS.—The pigeons keep their bill open, which is filled with mucous liquid; the tongue looks yellow, and the breathing is heavier than usual. The birds are uneasy, much less lively than before, do not show the least inclination for coupling, the mates sometimes even separating. Really moulting is nothing else but the natural feather-shedding, wherefore many fanciers do not regard it as a disease.

CAUSES.—These are to be found in the change of plumage, which usually occurs about the middle of June and lasts for a length of time.

TREATMENT.—Let the fowls be provided with strong and copious food, especially barley, and from time to time give them a dose of **Natrum muriaticum,** or **Aconite** when there are signs of fever, a dose morning and evening. With this treatment the pigeons will get safely over the moulting.

If this has not been complete *half-moulting*
sets in, which, if not attended to at once, may
have dangerous consequences. It is principally
known by the growth of the feathers in a wrong
direction, although this symptom is less dangerous than others. All such perversely growing
feathers have to be carefully plucked out, without, however, wounding the fowl. Besides this
the same treatment as in ordinary moulting
has to be observed.

10. CORYZA.

SYMPTOMS.—The pigeons are suffering from
a yellow mucus in their bill and mouth, which
is infectious. Besides this malady shows the
same symptoms as the above-mentioned "half-
moulting," and is usually a sequel of this. It
is among the most dangerous and contagious
pigeon-diseases, and often makes great havoc
among them, calling for prompt precautionary
measures.

CAUSES.—The most frequent cause of this
malignant disease is an irregular and not
natural feather-shedding.

TREATMENT.—First of all the sick fowls must be removed from the cotes and treated separately in order to avoid the infection of the rest. Then the bill has to be opened and the mucus which has gathered in the mouth be removed with a little lint-wiper moistened with water. By then giving a few doses daily of **Acidum sulph.** and **Mercur. viv.** in alternation, the evil will be completely removed.

11. INDIGESTION.

SYMPTOMS.—The bird manifests sadness, diminished appetite, perhaps even aversion to food. The tarrying evacuation consists of only half-digested or quite undigested food. Often there is even retching and vomiting.

CAUSES.—This complaint in pigeons is owing to the too hasty swallowing of great quantities of seeds, which then become crammed in the crop and stomach to such an extent that they cannot be digested and turn putrid, and are thus very apt to kill the fowl.

TREATMENT.—Confine the animal and reduce its feeding, or let it even fast for a while. Be-

sides give **Nux Vom.**, two doses a day, and if
the malady should not yield then **Dulcamara.**
For complete loss of appetite and constipation,
Bryonia, and if that is insufficient, **Arsenicum**
will afford relief.

12. WORMS.

SYMPTOMS.—As signs that a pigeon is affected
with worms we may recognize that its eyes look
blurred, watery and pale, that the fowl although
taking food falls away, and that the dung smells
very offensively. These signs, however, are
sometimes illusory and may as well be referred
to other causes. The only reliable sign of an
excess of worms in the bird is their being passed.

CAUSES.—There is in most cases some morbid
mucous secretion, disturbed digestion, etc., at
the bottom of this complaint.

TREATMENT.—If you observe worms passed
with the dung give several doses of **Cina,** the
main remedy for worms and all the troubles
they cause. Besides two doses of **Sulphur** given
every week for some length of time will be
useful.

All this care in the diseases of pigeons can, as will easily be understood, only be extended to precious and costly fancy pigeons, since it would be a hard matter to know the diseases of single birds in an extensive flock.

c. **A Glance at the Natural History of the Domestic Hen, her Diseases and their Homœopathic Cure.**

The domestic hen belongs to the seventh order of birds, the so-called Gallinaceæ. As their principal characteristics, are regarded the red comb on their head and the wattle hanging from the lower part of their bill.

We need scarcely remark that there are different species of hens. Their diversity lies in their size and external conformation. So there are large and small species, species with long and short legs, naked or feathered feet; there are hens with and without a tail, as well as there is great variety in the coloring of the plumage.

The cock or rooster, distinguished by his size, form, proud gait, beautiful plumage, and the spurs at his feet, which grow with his age is, quite superior to the hen. He has from time immemorial been praised as the symbol of watchfulness, courage and bellicose mood, while even at the present day his wide-awakeness makes him the clock of the farmer and in many cases his barometer. The hen, on the contrary, is the image of motherly care and tenderness.

As to food Nature herself has indicated to the hen a varied diet consisting of animal as well as vegetable substances, and her well-being depends in a marked degree from a judicious combination and supply of food of either kind. Among the vegetable food the hen is particularly fond of seeds, as barley, wheat, buckwheat, corn; neither will they refuse boiled potatoes, fresh clover, grass, salad, etc. Of the animal substances they prefer worms, little insects, etc.

The hen is most useful to man by her eggs, which form a most healthy and nourishing article of diet, and by her flesh, which is prepared and eaten in the most manifold form. Hen-breeding, therefore, offers the farmer considerable chances. At the same time, however, the hen is liable to a great many diseases, a knowledge of which is indispensable to the breeder who expects profit from his investment.

As soon as the hen falls sick the redness of her comb disappears, and this turns pale, flaggy, and takes a yellowish color. The feathers stand up and lose their gloss, the gait becomes slow. The bird is sad and cast-down, it separates from the rest of the healthy hens, who often persecute and bite it severely. Besides this every disease has its own characteristics, which we will try to specify here below.

1. SORE EYES.

SYMPTOMS.—This malady shows itself by ulceration and watering of the eyes, and at last causes the formation of pus-sores, which, without timely help, will lead to a fatal termination. At the same time the patients fall rapidly away, as all the liquids of the body are attracted towards the diseased parts.

CAUSES. Many breeders ascribe this malady to climatic causes; but it may as well be brought on by overheating, dust, damp stables, etc.

TREATMENT.—A few doses of **Aconite,** given daily, have always proved very efficient. **Belladonna** and **Euphrasia,** given same way, may also be useful if called for.

2. FRACTURED BONES.

SYMPTOMS.—We know them by the same signs as in pigeons, but they are much more frequent with hens than with the latter. The *causes* and the *remedies* to effect a cure are likewise the same, wherefore we refer the reader to what has been said above under this chapter. We may add, however, that the requisite splints may easily be made of the hollow stems of the common elder by scraping out its pith. We may repeat, too, that the patient, to protect it from being bitten or run over, ought to be secluded. The best way to do this is to assign it to a lay of soft straw or hay in a basket.

3. HERNIA.

(Protrusion of the laying-gut.)

SYMPTOMS.—The laying-gut of the hen especially of the larger species, is forced out during laying to such an extent that it does not recede, and thus is caused a hernia, which indeed can be reduced, but returns with every evacuation of dung.

CAUSES.—In most cases the protrusion of the laying-gut is caused by the repeated and unsuccessful efforts to expel an unusually large egg.

TREATMENT.—After noticing that a hen remains several hours upon the nest without success, if through examination the presence of an egg of uncommon size be ascertained, the simplest way is to tap the egg, in order to let the contents flow out, whereupon the shell will speedily follow. If the damage be already done the shortest procedure of course will be to kill the bird, if it is of no particular account to preserve its life. This may, however, be the case with a valuable and costly hen. The parts concerned have then to be bathed in tepid water or lukewarm milk, and the protruding gut rubbed with linseed oil and gently pressed back into the body. The protrusion is likely to recur several times, but will disappear at last, supposing the breeder bestows the requisite attention and care upon the fowl. Besides these manipulations the patient ought to be given **Aconite,** when there are signs of fever, or **Arnica mont.** One dose twice daily.

4. CONSUMPTION.

(Marasmus.)

SYMPTOMS.—This malady consists in an inflammation of the glands of the feathers near the rump, from which an oily fluid is secreted, which, if pressed out with the bill, serves to the healthy fowls to lubricate the plumage, and thus keep it pliable. If these glands become obstructed, inflammation, pain, fever, heat, and constipation set in. The sick birds are sitting about sad and morose, cease scratching the soil and feeding, and try to re-open the glands. If nature or art do not assist them they emaciate and die.

TREATMENT.—The hardened and obstructed glands have first to be touched by softening remedies, linseed oil, yolk of eggs, etc., and after having become soft to be opened with a very sharp knife; the degenerated lymph which has gathered there has to be pressed out, with due care, however, not to squeeze the suffering part, and then the wound to be bathed with **Arnica water.** If the inflammation has not been of too long standing already, the fowl will recover under this treatment. Internally **Aconite,** one dose every four hours, has to be used along with it.

5. DIARRHŒA.

SYMPTOMS.—The patient's dung is liquid, they fall away and cease laying.

CAUSES.—This complaint is caused by damp, cold weather, cold of the stomach, brooding in damp, cold stables, feeding on noxious berries, or too many ground worms, want of the sand or lime necessary for the digestion of hens.

TREATMENT.—The principal remedy is the same as mentioned for the diarrhœa of pigeons, **Ipecacuanha.** If faulty feeding be the exciting cause, **Arsenicum** has to be given. **Chamomilla** has also proved useful. One dose of the selected remedy every two hours.

6. WANT OF APPETITE.

SYMPTOMS.—Hens frequently lose their appetite if they eat too many worms and too few
seeds, or decayed and musty food.

CAUSES.—As just stated, or weak digestion.

TREATMENT.—In most cases the help of
nature is sufficient, the hens fasting. If they,
however, should become feeble and ill-humored,
and tardy in laying eggs, art must step in.
Arsenicum is availing in most cases, to be
changed for **Nux vomica** and **Dulcamara**
where there is a weak digestion. One dose of
either thrice a day.

7. TUMORS.

SYMPTOMS.—Every one will know a tumor at
the first aspect.

CAUSES.—They may be want of cleanliness,
external injuries and wounds, which, if neglected or falsely handled, will turn into tumors
or some inward disease of the animal.

TREATMENT.—Where the malady is accompanied by inflammation and heat **Aconite** has

to be given first hourly.   The tumor being ripe has to be opened, pressed out, and bathed with lukewarm water.

8. THE HEN-DISTEMPER.

SYMPTOMS.—The hens lose their bright looks, crouch about in corners, and die readily one after the other.   Closer examinations shows the skin around the anus to be colored a high-red with black spots.   This malady is contagious, and has a character similar to inflammation of the spleen.   (Milzbrand.)

CAUSES.—As this plague is mostly observed during hot and dry weather, the exciting causes are commonly attributed to atmospheric conditions.

TREATMENT.—Against this malignant distemper a lot of allopathic remedies may be found in the common books on hens.   None of them, however, will afford any real relief.   The principal remedy of Homœopathy is **Nux Vomica,** which has proved splendidly successful.   Give a dose three times a day.

The interest of the hen-breeder will, of course,

command him to immediately separate the diseased hens from the healthy, thus preventing all contact between them. The coop has also to be carefully cleaned by sweeping and washing, and frequently to be strewn with fresh sand. To fumigate it with Juniper-berries is likewise advisable.

9. WHITE COMB.

SYMPTOMS.—This disease consists in a gradual discoloring of the fleshy comb, which from below upwards becomes slowly covered with a whitish dust, which, after a while seizes the wattle also and then passes to the skin of the neck, the feathers of which die off by degrees, leaving the dead quills only. The malady having once reached this extent the hens will usually succumb to its rapid spread. It seems to be contagious too and most liable to befall the Cochin China hens.

CAUSES.—Dr. Gerlach, of Berlin, and other savants of note, have demonstrated this plague to have its origin in the formation of a vegetable parasite, a *fungus*.

TREATMENT.—Caution requires to separate a hen thus affected from the rest, providing for a healthy place, good nursing, and appropriate feeding. The treatment, in order to kill the morbific germ, has to commence by giving **Sulphur,** which may be followed by **Staphisagria,** which is the main remedy in most cases. A dose three times a day.

10. SWELLING OF THE HEAD.

SYMPTOMS.—The hen does not eat, lets her wings hang, and is sad and crouching about. The head swells very much and the fleshy parts indicate heat.

CAUSES.—This malady is caused by partaking of musty food or putrid water, as well as by a general disturbance of digestion.

TREATMENT.—Seize the sick fowl and wrap it up to its neck in a thick cloth, some old apron or such like. The hind part has to remain uncovered to prevent its becoming soiled. Then the head has to be wrapped up in wet linen rags, and this dressing renewed frequently. Inwardly **Belladonna** and **Arseni-**

cum may be given in alternation, substituting **Bryonia** if the swelling of the head be very hot and tense.   A dose once every hour.

11. CONVULSIONS.

SYMPTOMS.—The birds are attacked by jerks, fall down suddenly, flop their wings, and stretch out their legs spasmodically; even death may set in in a moment.

CAUSES.—Chasing the fowl about, especially during great heat, bad food, or wet and cold weather may bring on this disease.

TREATMENT.—Take the sick animal immediately to the hydrant and pour cold water on it.   This is the simplest and best remedy.   It must then be secluded and provided with soft food, as boiled barley, and in summer salad and cabbage.   Heating food, among which is rye, must be avoided.

Homœopathy would, after a few doses of **Bryonia,** prescribe one dose of **Nux vomica** daily.   **Belladonna, Hyoscyamus** and **Mercurius vivus** have in many cases also proved effective remedies.   One dose twice a day.

12. ITCH.

SYMPTOMS.—The presence of this complaint may be known by the constant scratching and biting of the fowls, who at the same time become droopy and lose their feathers. An examination shows that the body of the patient is more or less covered with an itchlike eruption, which is particularly frequent and more coarse than fine on its back. This malady being infectious, great precaution has to be taken.

CAUSES.—Filthiness of the coops, foul, decayed food, want of a chance for bathing, and of clean fresh water for drinking. The complaint is more frequent in summer than in winter.

TREATMENT.—First of all the exciting causes have to be removed. Better nursing, more wholesome food and proper clean roosts will greatly assist a cure. The internal treatment has to commence with **Sulphur,** one dose given daily for three days. This has to be followed by **Staphisagria,** which is the principal rem-

edy in most cases. If the eruption is vesicular,
containing a yellowish fluid, which hardens
into scurfs, give **Dulcamara. Sulphur** is
then given to complete the cure.

13. SWELLED CROP.

SYMPTOMS.—The hen is sad, ruffles her feathers, throws her head about, the crop and sometimes even the head are swollen.

CAUSES.—Swelling of the crop is produced
by an excess of food, and can easily be ascertained by examining the crop, which is extended to such a degree that it would seem
ready to burst.

TREATMENT.—Sometimes the hens help themselves, standing quietly up in an erect position,
the head stretched upwards, and thus waiting
till digestion is done. This may be assisted
and promoted by a few doses daily of **Arsenicum** and **Nux vomica,** in alternation, every
two hours. Should there, however, be no improvement, the crop has to be opened with a
sharp penknife a little sideways where the
crammed food can be felt, the cut to be about

one inch long, whereupon, using gentle pressure, the food has to be lifted out with the finger. The wound has then to be sewed up with silk thread and spread with unsalted butter or linseed oil. Under this treatment the wound soon heals and scars over, and the hen is saved. The hen has to be kept apart from the rest in a moderately warm place, and fed with soft substances, as soaked bread, chopped grass, salad, cabbage, etc.

If the swelling of the crop be neglected it is very liable to pass into *induration* and *tearing of the crop*, which call for the same operation as we have described.

14. LICE.

SYMPTOMS.—The hens have often greatly to suffer from fleas and lice. They then waste away, lay very little, and grow constantly weaker. If such a lean hen is examined the lice can often be found in heaps.

CAUSES.—Filthiness of the hen-coops, want of baths of sand or dust. Infection too may cause the complaint.

TREATMENT.—First of all let the lousy hen be separated from the others, in order to prevent the spreading of the disease. The ailing fowl must then be washed with a decoction of **Absinthium** (wormwood). **Oil of fennel** dropped on head and neck will also drive the lice out. To the hens suffering from them **Absinthium** and **Sulphur,** and where there is great debility, **China** will have to be given. A dose twice a day.

In order to clean the coop from lice this ought to be white-washed. To give the walls a coat of "soluble-glass" or Chloride of lime is also available. If the green twigs of alder are put into the coop and removed again the next day, the lice which cling to the twigs will likewise be got rid of. As precautionary measures it may strongly be recommended to cleanse the hen-coop every week, and to strew the floor with sand, sawdust, or powder of ferns, etc.

Besides keeping the coop and the fowls clean, they should never be without sand, dry dirt, or ashes, for the hens bathe in these and fill their plumage with them. By virtue of the muscles of their skin they give a shaking movement to the plumage, by which the lice are thrust off with the sand, to be eagerly picked up and eaten by the hen.

15. MOULTING.

The SYMPTOMS and CAUSES have already been stated when we treated of the moulting of the pigeons.

The TREATMENT is limited to taking good care that the fowls be kept warm and have no want of good vigorous food and pure drink. Many breeders, as has also been mentioned before, think that moulting is no disease at all.

16. THE HUMID OR BLACK DISEASE.

(Sweet-malady).

SYMPTOMS.—This malady befalls only hens hatching in damp, musty coops, and its sign is, that the featherless parts under the wings become moist, clammy, and blackish-looking.

CAUSES.—The hen is too eager to hatch. She does not bathe in sand and air her plumage.

TREATMENT.—The affected parts have to be

3

washed with lukewarm water, and powdered with flour, and the hen made to air herself every day.

17. "PIP."

SYMPTOMS.—The sick hens are sad, do not scratch, like to hide, sit with open bill and ruffled feathers, they are bloated, snore, and utter from time to time a sound like "zip, zip." Their nostrils become blocked by tough mucous secretions, the comb hangs sideways, loose and withered, and of pale, yellowish appearance, the skin of the nether side of the tongue is hardened. The crop is mostly empty on account of the difficult swallowing.

CAUSES.—This disease originates in an inflammation of the throat, caused by sudden change of the weather, getting cold, particularly in the feet; by hot food, want of green nourishment and insects, water which has become stale or foul, especially in wooden troughs.

TREATMENT.—According to some old traditional usage this malady is treated by pushing, with the index-finger, the throat of the patient gently backwards, and after the tongue has been pulled out sideways, loosening with a sharp penknife the hardened nether skin of it and peeling it towards the tip. This proceeding, however, which has been continued from generation to generation, is based on a perfectly erroneous opinion. It is an error to assume that the disease consists in the hardening of the tongue, whilst this is only a symptom of a malady, which, if the latter itself be removed, will likewise disappear.

TREATMENT.—"Pip," considering its true exciting causes, is best cured by separating the sick fowl for some days from the rest, and feeding it on soft food, as bread soaked in milk, green vegetables, potatoes, etc. Inwardly **Spongia**, a dose three times a day, may be administered. In many cases the first dose of this remedy has effected a cure.

18. SMALL-POX.

SYMPTOMS.—The birds afflicted with this disease are melancholy and weak. By close examination small tumors (pustules) of the size of a pea are discovered under the wings, on the belly and the inside of the shanks.

The CAUSES and the TREATMENT are the same as of the like disease of pigeons.

19. CONTUSIONS.

These are easily recognized. The treatment consists in bathing with diluted tincture of **Arnica.**

20. CATARRH OF THE NOSE.

(Coryza.)

SYMPTOMS.—The patient sneezes often, snores, the eyes water, and some slimy fluid is flowing from the nose.

CAUSES.—This malady is caused by catching cold in continuous wet weather during summer. Sudden change of the latter can also produce it.

TREATMENT.—As a rule nasal catarrh is a light complaint, and its removal effected by keeping the animal warm and feeding it on soft food. The duration is, however, much shortened by a few doses of **Mercur. vivus,** and in case the nasal slime should thicken, and the eyes be blurred, of **Euphrasia.** For a feverish condition give **Aconite.** Give the selected remedy thrice daily.

21. POISONING.

Poisoning cannot properly be called a disease, but may become the cause of sudden deaths. In most cases, indeed, the natural instinct of the animal will preclude their eating noxious plants, herbs, seeds, etc., but this is not without exceptions. The hens, for instance, will eagerly eat bitter almonds, which are decidedly fatal to them. Poisons and substances of this kind, if swallowed by hens, will usually act so rapidly and powerfully that all assistance is useless.

22. CONSTIPATION.

SYMPTOMS.—The hen is restless and cannot evacuate, although she makes frequent efforts to do it; or the dung is only passed in very small quantity, and exceeding hard and dark.

CAUSES.—This complaint is caused by too long-continued, dry, hot food, as barley, oats, rye, hempseeds, and other kernels, accompanied with want of pure water and green vegetables. It occurs, therefore, especially among hens kept in narrow yards. But it may also be caused by cold.

TREATMENT.—It has to be commenced in all cases with a dose of **Aconite,** which has to be repeated as long as the restlessness of the hen increases. Then give **Nux vom.,** which usually removes the complaint. If costiveness is the sequel of a cold, **Bryonia** is the proper remedy.

23. DROPSY.

SYMPTOMS.—This malady manifests itself by swelling of the abdomen. The hen continues to eat, but walks about with ruffled plumage.

CAUSES.—As a rule this disease befalls only very old hens, particularly such that have ceased laying but have greatly fattened, the fat at last dissolving into watery elements. As this involves a general decomposition of the fluids of the whole body, it is a hard task to cure this disease.

TREATMENT.—Against any dropsical condition **China** and **Arsenicum,** administered in proper dose and alternately, are the most available remedies, always supposing, however, that the disease has not reached too far. Besides those named, **Lycopodium** is a most efficacious remedy, if there be considerable external dropsical swelling.

24. INJURIES.

Wounds and all kinds of external injuries are so easy to ascertain that it is superfluous to state particular symptoms.

CAUSES.—These are very manifold. Hens and cocks, for instance, frequently hurt their combs and eyes by strolling among the thorn-bushes, or they are injured by the thrusting at them of pointed objects, etc.

TREATMENT.—The first thing to be done for the cure of wounds is to clean them of foreign bodies that may have penetrated, and then, if need be, a good dressing. Wash the wound, therefore, with water, and try by other appropriate manipulation to remove anything foreign. In order to stop much bleeding, cold water or mixed with vinegar is the best means.

To the wound itself **Arnica** is applied externally, which may also be given internally if the wound be a deep one. If traumatic fever supervene give **Aconite,** alternating with **Arnica.**

If maggots appear in a wound in summer, they ought to be wiped off and the spot spread with tar to prevent the further tainting with fly-blow.

25. WORMS.

SYMPTOMS.—The hens waste visibly away in spite of cleanliness, good nursing, and proper feeding, while their appetite is not in the least impaired. This is a sure sign that they are suffering from worms.

CAUSES and TREATMENT are the same as for the like affection of the pigeons.

26. GOUT.

SYMPTOMS.—The legs and claws of the hens swell, they grow quite stiff, and cannot sustain themselves any more on their perches. They lose their appetite and their sprightliness.

CAUSES.—Severe cold in winter as well as ..lthiness, and the walking on paved yards may produce this malady.

TREATMENT.—The first thing to be provided for is warmth. Then the legs and claws of the patient ought to be rubbed in the morning with brandy, containing a solution of kitchen-salt. Half an hour later the suffering swelled parts have to be spread with sweet (unsalted) butter or melted mutton-tallow. This will in a short time effect a cure.

d. A Glance at the Natural History of the Turkeys. Their Diseases and Homœopathic Treatment.

The turkey belongs to the order of gallinaceous fowls, and to the family of hens properly. It is distinguished from the other kinds of hens by a short strong bill, a curved upper jaw, which is topped by a fleshy gland, and by a wattle reaching from the head and neck down to the breast, of a wrinkled and warty structure. Its proportionally very small, unfeathered, and almost bald head, as well as one-half of the neck, is covered with a naked, bluish skin, beset with many partly red partly whitish wart-like excrescences. The long root of the powerful legs of the male is armed with a weak spur. They have twenty-eight wing-feathers in each wing, and eighteen such in the tail, which they are able to erect as well as to spread like a wheel.

The male turkey, which during the first days after his birth is smaller than the female, soon overtakes this and grows stronger and bigger. The turkey-hen is distinguished from the cock principally by the want of the spur at the claw, and of the hair-tuft on the breast; the fleshy gland on the upper part of the bill is less dark-colored and shorter, and not extensible. The other fleshy glandular parts of the head are also of a paler shade and less protruding. Besides, the turkey-hen at the time of puberty becomes smaller and more graceful. Although being likewise provided with a double-tail, she can neither erect nor spread it.

Concerning the habits of the turkey, we prefer to quote some extracts from a good work on fowl-breeding by H. Gauss:

"The very remarkable habit of the turkey-cock, to spread his tail wheel-like, does not show itself in rest, but only in moments of pride, and makes this beautiful proud fowl a real decoration, and the imposing ruler of the barn-yard. It develops in the spring of the second year of his life, along with the complete formation of the body and sexual maturity. Instigated by love, scorn, or jealousy, he then blows up his throat, the head and rest of the neck swell also, the fleshy parts take a lively and glossy red, the cubiform flesh gland on the upper jaw of the bill extends and enlarges to the size of about two inches, the plumage of the cock stands on end, the tail rises fan-shaped, while the spread wings drag on the ground. At the same time the feathers are set into motion by muscular contraction, and their rubbing against each other produces a dull rustling sound. This is accompanied by a piercing cry, which is called gobbling, which he repeats if one whistles or mocks him, or excites his anger. The scorn gets highest if he sees some red cloth; he flies at it furiously, picks it with his bill and tries to remove it.

"The tame turkey eats every animal or vegetable substance that may serve to nourish, but prefers insects and grain. Grass and the leaves of other plants they only take when they find nothing else. They are even fond of raw or boiled flesh, and hunt smaller animals, as frogs, mice, and lizards, which they tear and swallow with delight.

"The turkey is useful to man by his flesh, which is prepared in various styles, and is a considerable article of traffic. Many thousand cocks and hens are shipped to England every year, and the fattened and truffled turkeys of Limoges, Brives-la-Gaillarde, Kahoos, Perigneux, and other French towns, are exported to all countries of Europe.

"The turkey, if once grown up, is less subject to disease, but in his youth he is liable to a great many afflictions and ailments. The turkey-breeder, therefore, in order to escape the loss of whole breeds, has to bestow particular attention and care to the turkey-chicken. The cause of this is to be found in the soft nature and tender constitution of the chickens, which are easily destroyed by anything injurious occurring to them. Cold and dampness are their mortal enemies, therefore they must be kept dry and warm. If they have to sit in their coops on a cold stony ground, they take cold in the legs and become lame, or die of it, or are crippled and will never thrive. For the same reason they cannot stand thaw, hoar-frost, and snow. Great heat is alike deleterious to them; the sun scorches their back, causing the feathers to fall off and gradually to die away."

1. VESICLES UNDER THE TONGUE.

SYMPTOMS.—Very often vesicles appear under the tongue of the turkeys, which become very troublesome and deprive them of all their sprightliness.

TREATMENT.—The vesicles are carefully opened with a needle, the fluid they contain is pressed out, and the sore moistened with some **Arnica water.**

2. DIARRHŒA.

SYMPTOMS well known.

CAUSES.—The turkeys are liable to contract this complaint by eating too many snails, rain-worms, etc., of which they are particularly fond, or by partaking of wet grass, or by the excessive use of sweet and soft dough-food.

TREATMENT.—Such a diarrhœa must not be made light of too long, for it tells greatly on the tender turkey, debilitates, and at last kills it. Green and wet food must be shunned altogether, and **Ipecac.**, **Arsenicum**, or **Chamomilla** be given internally. A dose of the selected remedy about every two hours.

3. GOUT.

SYMPTOMS.—The legs swell and become stiff, the fowls cannot tread on them, and have an uncertain gait.

CAUSE.—Taking cold.

TREATMENT.—Let the patient be brought to a warm and dry place, wrapped up in warm wadding, wool or tow, giving **Bryonia** inwardly.

4. APPEARING OF THE WART-SKIN.

SYMPTOMS.—If the turkey-chickens at the age of two months lose their appetite and walk about sad, we may be sure that the wart-skin will soon make its appearance. This is a crisis which destroys many of them.

CAUSE.—The appearing of the wart-skin.

TREATMENT.—Feed the chickens on nourishing but easily digestible food, give them invigorating drinks and expose them to warmth, whether it be the heat of the sun or artificial. Internally give **China** or **Chamomilla** twice a day.

5. COUGH.

SYMPTOMS.—The fowls suffer from a hollow-sounding cough, and are continually threatened with suffocation.

CAUSES.—There is an accumulation of small red worms in the windpipe of the fowl and its branches, which cause this cough.

TREATMENT.—**Camphora** has proved available in many cases, as likewise **Absinthium.** Give a dose three times a day.

6. MOULTING.

SYMPTOMS and CAUSES the same as with other fowls.

The turkeys too, as many other birds, are subject to this periodical weakness. They are then sad and drooping, the feathers erect themselves, they shake them frequently and pull them with the bill, in order to make them fall out; the fowls do not eat, and many die.

TREATMENT.—They ought to be kept warm, taken early to their roost and not let out too early, to protect them from cold and dampness. They ought to be fed during this period on hemp and millet. Besides give **Natrum muriat.,** and if fever be present **Aconite.** A dose night and morning.

7. SMALL-POX.

SYMPTOMS.—Small-pox appears in the form of little vesicles around the bill and inside of it, on the bare parts of the body, and under the wings and thighs, and on the warts. The small-pox is a disease as contagious as fatal, against which not much can be done, wherefore many breeders kill the fowls' as soon as there is the least sign of the disease.

CAUSES.—The most frequent exciting cause may be found in unfavorable atmospheric conditions. As another one we may regard its contagiousness, since the exhalation of one single bird may infect the whole flock.

TREATMENT.—Let the patient be separated from the healthy fowls, and give it **Arsenicum** and **Rhus toxicod.** in alternation every two or three hours.

8. CORYZA.

SYMPTOMS.—When this malady commences the fowl feels uneasy, it begins to tremble, an acrid, slimy fluid flows from the nostrils, and the lustre of the eyes is almost wholly gone

CAUSES.—This malignant and contagious disease is produced by an inward disorder, in many cases a sequel of an abnormal course of moulting.

TREATMENT.—First of all, the sick turkey has to be separated from the healthy hens, and be treated apart, in order to preclude all further contagion. The gathering mucus is removed with a small lint rag, which has been moistened in water. The principal remedies are **Acid. Sulphur.** and **Mercurius solub.** A dose of each three times a day.

9. DEBILITY.

Debility or the so-called *lymphatic condition* of the turkey-chickens has to be counteracted by invigorating food and drinks, or by very careful nursing generally. To give them strong claret and bread soaked in wine is much to be recommended.

10. THE TUMOR

Of the turkey has to be treated in the same way as the wasting of the hens. We refer, therefore, the reader to that chapter.

11. VERMIN

In turkeys have to be got rid of in the same manner as in hens, geese, and other domestic poultry.

12. CONSTIPATION

Occurs principally in too lascivious males. The treatment is the same as for hens.

e. A Glance at the Natural History of the Goose, her Diseases and their Homœopathic Treatment.

The goose belongs to the order of the web-footed and swimming birds, and the family of the so-called *Blattzahner.* Our domestic goose has her origin from the wild goose. Her domestication, however, has not left any trace of her wild qualities; she seems even to have forgotten the sweetness of her former liberty, never trying to regain it, as the duck does, for instance. She lives in peace with the rest of the poultry, causing no disorder or strife among them. Through breeding she has grown much larger and stronger than the wild goose. Her wings, on the contrary, are less strong and stiff, and the color of her plumage, too, has changed entirely from that of her wild sister. Her constant cackling and excessive chattering has

caused our ancestors to give saucy chatter-boxes
the attribute of a saucy goose. We also hear
often the expression "stupid goose," although
the goose is by no means as stupid as she looks,
and has already shown many traits of prudence
and understanding.

The gait of the goose is slow and awkward,
which is caused by the relative position of the
legs to the rest of the body. The gander is
larger and stronger than the goose, has higher
legs, a stronger and longer neck, and a deeper
voice. He is also devoid of the so-called "lay-
ing-belly" of the older females. During the
time he has to lead the chickens he is apt to
become rather dangerous; for he flies at his
putative enemy with enraged cries and spark-
ling eyes, bites and beats him with his wings.
Small children must be carefully protected from
such attacks.

On the rump the geese have a group of
feathers, which are but very small and lie
hidden by the other feathers. They surround
the efferent duct of the adipose glands, which
lie sideways of the os sacrum (sacred bone).

These feathers always contain a good deal of fat, with which the goose lubricates the rest of her plumage, which accounts for the water not sticking to it but quickly running off.

Although the goose is not over quaint and whimsical in regard to her food, she neverthe-less prefers the sweet oats and barley, and among the green vegetables, the clover, grass, vetches, etc. She likes also mashed potatoes, carrots, white beets, and the leaves and trunks of cabbages, as well as bread, bran, grits, etc. She needs a copious supply of pure water.

The goose is useful to man by her flesh, which is prepared in the most various styles; that of the trunk furnishing a delicious roast; or smoked and pickled the well-known goose-breast, cele-brated as a dainty, is especially shipped from Pomerania all over the world. The grease is a substitute for the costlier butter, and was highly prized by the ancients as a nerve-invigorator and cosmetic. Even in the middle ages it used to be a favorite remedy in a variety of diseases. The feathers serve for stuffing our beds and for artistic purposes, and furnish the best writing material, scarcely even to be superseded by the steel pen.

The goose, as a rule, enjoys durable health, and is less liable to disorders than other fowls, except the duck. There are, however, single maladies, as, for instance, the spleen-distemper, which may befall whole flocks and thus become very destructive. To know and to cure them is the task of the breeder. We treat, therefore, below of the single diseases, referring, however, to those of the hens and their treatment where both of them generally coincide.

1. VESICLES.

SYMPTOMS.—There appear on the neck of the geese little vesicles, containing a contagious fluid.

CAUSES.—Some morbific germ dormant in the animal, which if the favorable conditions are given, develops into that malady.

TREATMENT.—As soon as the complaint is noticed the sick individual ought to be confined alone and **Sulphur** given in repeated doses. For food, chopped salad is appropriate. If good nursing be added to this, paying especial attention to cleanliness, the malady will soon disappear.

2. STAGGERS.

SYMPTOMS.—Under the influence of this disorder the geese let their wings hang down, stretch the neck, wag the head, shake themselves constantly, do not eat, and turn round in a circle until they are seized by vertigo, of which, unless speedy help is given, they die in a few moments.

CAUSES.—This malady is caused by too strong influx of blood to the brain, or by the presence of worms in the ears or nostrils. .Sometimes a heavy blow, or something the like, on the head, causing concussion of the brain, or even a fracture of the skull, may bring on similar affections.

TREATMENT.—To effect a cure, the different exciting causes ought to be taken into account. If the disease is caused by rush of blood, the vein under the web has to be opened with a sharp penknife. This is easily found out, and in this way blood may be drawn. If worms be the cause give some doses of **Belladonna** every day. If injuries brought on the complaint let some repeated doses of **Aconitum** precede **Belladonna.** The head may also be bathed with **Arnica.**

3. DIARRHŒA.

This malady affects young geese in preference. The symptoms, causes, and treatment are the same as with chickens, where we refer the reader.

4. TUMORS

In geese are cured in the same manner as in chickens.

5. CONSTIPATION

Or costiveness of geese requires the same treatment as those of chickens.

6. FRACTURE OF BONES.

The symptoms and causes have been given under the same head for pigeons and chickens.

TREATMENT.—In many cases it is sufficient to confine the goose in a narrow place, where she cannot roost on any stone. It is not necessary to splint the fractured part, as quietude alone suffices to effect a cure. But the fowl must be provided with plenty of wholesome food and fresh water.

7. LICE.

Lice are driven from the geese by the same remedies as from the chickens. Cleanliness is often sufficient to get rid of the vermin.

8. MOULTING.

The geese too are subject to moulting, and the same treatment has to be followed as for the moulting of other poultry.

9. SPLEEN-DISTEMPER.

SYMPTOMS.—The head of the goose is greatly bloated, she vomits a greenish, tough, sour-smelling phlegm, which spins from the bill to the ground in long strings; feet and wings, instead of their red covering, appear yellowish and flappy. This disorder usually sets in in June and July, and often destroys whole flocks.

4

Causes.—Many breeders look for the exciting cause of this malady in atmospheric conditions.

Treatment.—Separate the sick goose from the other healthy fowls, feed her on light food with green vegetables, and give internally **Aconite, Arsenicum** and **Nitric acidum**. A dose of either two or three times a day.

10. "PIP."

The "pip" of the geese is treated and cured in the same way as that of the chickens. See page 58.

11. WHITE DYSENTERY.

Sypmtoms.—The disease commences with sadness of the fowls. They let their heads and wings hang, their appetite is diminished and disappears completely with the increase of the malady. The patients grow weak and lose their strength, so much so that they are scarcely able to stand up, but frequently collapse. Their breathing becomes hurried; the dung, at the commencement yet consistent, grows soft, of a chalk-color, and gradually turns liquid. The body at last assumes a bluish appearance, and the patients die in a few hours.

CAUSES.—The exciting causes are principally bad food, filthy coops, browsing on swampy, mossy ground, and damp, cold temperature. The course of the disease is rapid, and decided in from three to four days. With those fowls with which on the fourth day a critical decision sets in, the malady may be considered as removed on the sixth day.

TREATMENT.—As to the prevention and treatment of this disease generally, care has to be taken in all cases to bestow very careful nursing and feeding on the animals. They must not be driven to the pasture nor allowed to go into the water, and their coops be provided with fresh dry straw at least twice a day. Inwardly, at first, a few doses of **Aconite** may be given to counteract the inflammatory condition of the digestive organs. This has to be followed by **Arsenicum,** after which **Mercurius corrosivus, Rheum,** and **Chamomilla** may be appropriate. A dose of either every two or three hours.

12. APOPLEXY.

SYMPTOMS.—The goose is seized with vertigo, staggers, and suddenly breaks down; her breathing is heavy, short, and more or less snoring; the eyes are highly injected, protruding and staring; the circulation of the blood is disordered and checked.

CAUSES.—These are to be found in overfeeding the fowls with hot and too nutritious food, wherefore this malady often befalls geese while being fattened.

TREATMENT.—First give some doses of **Aconite,** then cold douches and, if necessary, cold compresses on the head. The recovery, however, of such a patient remains doubtful.

13. RATTLE.

SYMPTOMS.—The geese are sad, grow weak and feeble, stretch their neck upward, keep their bills agape, and cannot inspire through their nostrils. They shake their heads also, trying to get rid of the matter sticking in their noses. They frequently utter a rattling sound with this.

CAUSES.—The most frequent cause of this distemper is catching cold.

TREATMENT.—A small hen-feather has to be passed into the nostrils of the sick geese, in order to open a canal to the checked fluid. This operation has to be repeated on the second and third day. The bill has repeatedly to be moistened with lukewarm water, and the fowls must be kept warm and clean. For a drink give them perfectly pure water, and feed them on light food.

14. ULCERATION OF THE RUMP.

SYMPTOMS.—The goose is sad, lets her head hang, her sleep is disturbed and the bowels costive. The plumage becomes shaggy, and a tumor forms above the rump.

CAUSES.—This complaint is mostly produced by unclean water and filth in the coop.

TREATMENT.—Try to soften the tumor, and then cut it open with a sharp penknife, press the pus out with the fingers and wash the wound with watery vinegar, etc. **Mercurius v.** and **Hepar sulph.** may be given, a dose of each, twice a day. During the treatment the fowl has to be put under a refreshing diet, as green salad, barley, bran, and similar soft foodstuffs.

15. MORTALITY OF YOUNG GEESE.

During the months of June and July, that is, at the time when the goslings grow their quills, they are usually liable to great mortality. It is necessary to try to prevent this by all means, lest the breeder may experience great losses.

CAUSES.—The most common of these is, that the proper care in raising the young geese is neglected. If they are poorly provided for during the time of growing the quills, not well fed at home, so that they have to content themselves with the spare and poor food they find in the pasture, they must become weak and sick.

TREATMENT.—Considering that the fowls require strength to grow their quills, the breeder must pay particular attention to give them copious and strong food during that time.

Therefore, if you wish to protect the young geese from those morbid conditions let them be copiously fed before they are driven to the pasture, and do not neglect to do the same after their return from it. Under this treatment the mortality among the goslings will be considerably diminished.

16. POISONING.

SYMPTOMS.—A poisoning of the fowls may be suspected, if some of them fall suddenly and unexpectedly sick, betray anxiety, show trembling of the limbs and twitches, if diarrhœa and tenesmus set in, and there are pains in the stomach and intestines. Soon inflammation and mortification intervene, and unless speedy assistance be given death ensues.

CAUSES.—*Poisonous substances,* for the goose as such are considered: the seeds of petroselinum (parsley), the leaves of the poppy, the hemlock, the deadly nightshade (Bellad.), for which the geese have a great fancy, the seeds of the henbane (Hyosc. niger), the thorn-apple (Stramonium datura), etc. It is very frequent to see geese and other fowls miserably perish from having picked up the stray seeds of the named poisonous plants.

TREATMENT.—According to the substance swallowed the remedies to be given will be **Coffea cruda, Nux vom., Chamomilla, China, Camphora, Arsenicum,** etc. A cure may be effected too by tickling the fauces of the fowl with a feather, and thus causing it to throw up the poisonous substance. This will prove available in many cases.

17. WOUNDS

In geese require the same treatment as mentioned above for chickens.

f. A Glance at the Natural History of the Duck, Its Diseases and their Homœopathic Treatment.

Our tame duck, belonging to the web-footed birds, evidently originates from the wild duck, which is proved by the frequently occurring copulation between domestic and wild ducks in the neighborhood of large ponds or lakes. There is the other fact, too, that some originally wild species of this fowl have become domesticated and propagated, thus forming quite new tame races. In this way the Muscovy duck, by her plumage, flesh, and easy breeding, has become one of the most useful and widest spread species in the New World.

The figure of the duck is the same everywhere, and its size only differs on account of the care and feeding bestowed on it in its youth. Their waddling gait is owing to their feet, which at the same time serve them as paddles in swimming, being situated rather far backwards. The male, called drake, differs from the female by his larger size and a tuft of feathers above the crest, which stand one behind the other and are inclined forward.

The duck is an exceedingly voracious animal, swallowing everything which seems to it in the least palatable. To the fish ponds it becomes dangerous by swallowing little fishes and their spawn. But it does not refuse either snails or ground-worms, and even delights in

the intestines of butchered animals. Besides it
eats with pleasure kernels, particularly oats
and barley, as well as it feeds on duckweed,
ground carrots, soaked bran, etc.

The duck is useful to man by its flesh, which
gives a delicious roast, by its eggs, which are
highly palatable, and of which a much larger
number is laid than by the hen, and by its
feathers. The latter, although inferior to the
goose-feathers in regard to elasticity, are nevertheless very serviceable for the stuffing of
nether-beds and pillows.

The ducks, if we except the convulsions to
which the young ones are subject during the
first weeks of their life, are comparatively free
from diseases, which may principally be owing
to their excellent digestive powers and their
constant drinking, by which every undigesti-
ble or else noxious substance is eliminated.
The following maladies, however, may be mentioned here, for which the treatment in general
is that prescribed for geese diseases.

1. VESICLES.

Treatment the same as in geese.

2. VERTIGO.

This complaint too requires the same treatment as in geese.

3. COSTIVENESS AND CONSTIPATION.

Compare what has been said under this caption above.

4. FRACTURE OF BONES.

Compare the chapter of fracture of bones in other poultry.

5. CONVULSIONS.

Symptoms. — The ducks are attacked by twitches, stretch their legs in a spasmodic way, or draw them up again, flap their wings and fall down. This malady mostly befalls young and weakly ducks.

Causes.—Damp, cold temperature, poor food, filthiness of the coops, etc.

TREATMENT.—In most cases the complaint will be removed by pouring cold water on the animal.

Besides this, **Belladonna** or **Nux Vomica** may be given internally.   A few doses of either daily.

6. MOULTING;   7. DYSENTERY;   8. TUMORS;   9. VERMIN;   10. WOUNDS AND INJURIES.

Are attended to in ducks in the same manner as has been stated in detail for the rest of the poultry.